Ecosystems

© 2015 OnBoard Academics, Inc
Portsmouth, NH
800-596-3175
www.onboardacademics.com
ISBN: 978-1-63096-055-1

ALL RIGHTS RESERVED. This book contains material protected under International and Federal Copyright Laws and Treaties. Any unauthorized reprint or use of this material is prohibited. No part of this book may be reproduced or transmitted in any form or by any means, electronic or mechanical, including photocopying, reprinting, recording, or by any information storage and retrieval system without expressed written permission from the author / publisher.

OnBoard Academic's books are specifically designed to be used as printed workbooks or as on-screen instruction. Each page offers focused exercises and students quickly master topics with enough proficiency to move on to the next level.

OnBoard Academic's lessons are used in over 25,000 classrooms to rave reviews. Our lessons are aligned to the most recent governmental standards and are updated from time to time as standards change. Correlation documents are located on our website. Our lessons are created, edited and evaluated by educators to ensure top quality and real life success.

Interactive lessons for digital whiteboards, mobile devices, and PCs are available at www.onboardacademics.com. These interactive lessons make great additions to our books.

You can always reach us at customerservice@onboardacademics.com.

Food Chain

©2015 OnBoard Academics, Inc.   www.onboardacademics.com

What is happening here?

_____

_____

_____

_____

_____

_____

_____

_____

_____

_____

_____

_____

Food energy is being transferred from one living thing to another.

©2015 OnBoard Academics, Inc.   www.onboardacademics.com

A food chain shows the order of energy flow through organisms that live in the same environment which we call a biome.

All food chains start with the sun which provides the energy that plants use to make their own food. In this food chain the plants, the grass, is then eaten by a grasshopper. This is how the grasshopper gets energy. The grasshopper is then eaten by a frog. The frog is then eaten by a snake. Finally, the snake is eaten by a hawk. This order of feeding relationship or energy flow is called a food chain.

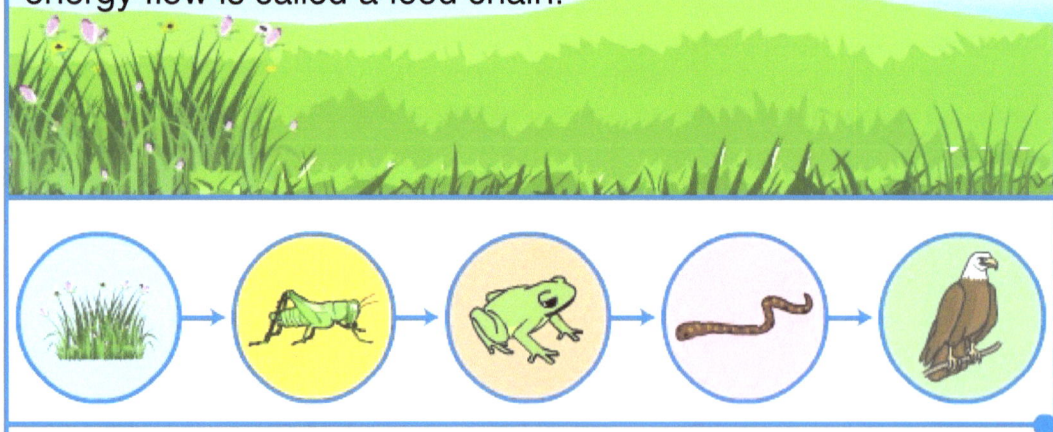

Roles in a Food Chain

> **It's easy to identify the role of a consumer in a food chain if you remember that food chains always begin with a producer, and that primary means first, secondary means second, tertiary means third, and quaternary means fourth.**

The living thing at the start of the food chain that makes its own food using the sun is called the producer.

producer

Living things that eat other living things are called consumers. In this food chain the grasshopper is a primary consumer meaning it's the first consumer in the food chain

primary consumer

The frog eats the grasshopper and is the secondary consumer. This means it is the second consumer in the food chain.

secondary consumer

Animals that feed on secondary consumers, in this case the snake, are called tertiary consumers. Tertiary means third.

tertiary consumer

Animals that feed on tertiary consumers, in this case the hawk, are called quaternary consumers. Quaternary means fourth.

quaternary consumer

Identify the role of each living thing in this food chain.

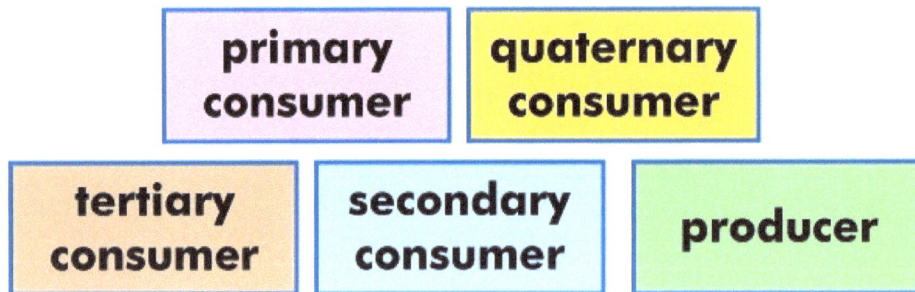

primary consumer   quaternary consumer

tertiary consumer   secondary consumer   producer

A food chain always starts with a producer, but the number of consumers can vary. This food chain shows only two consumers. Food chains rarely go above quaternary consumers (the level above tertiary).

Create a grassland biome food chain with the pictures and labels below.

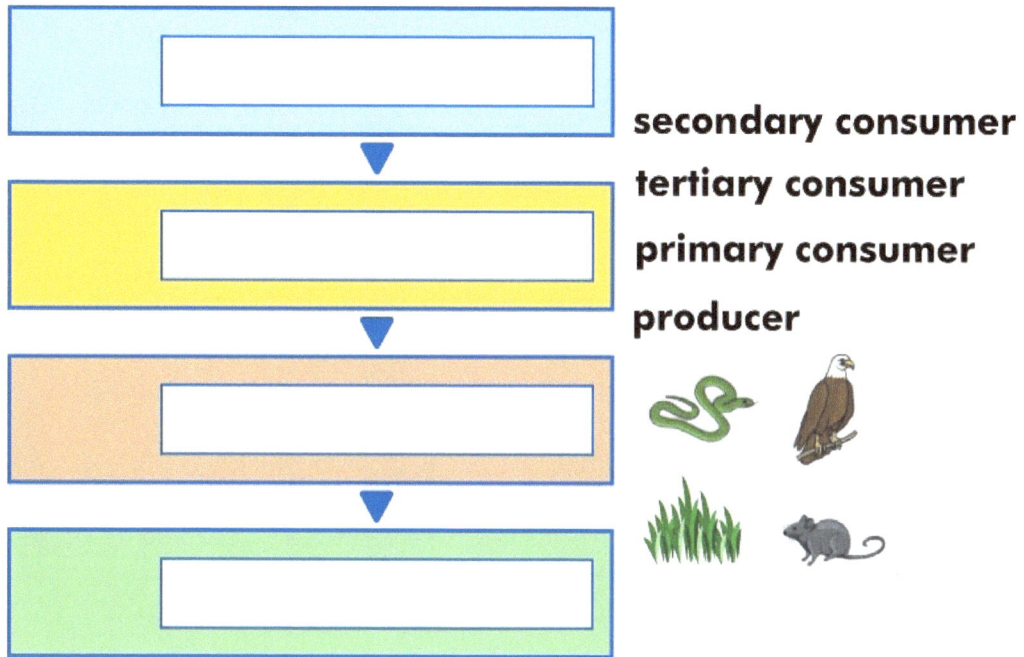

secondary consumer

tertiary consumer

primary consumer

producer

Create an ocean biome food chain using the information below.

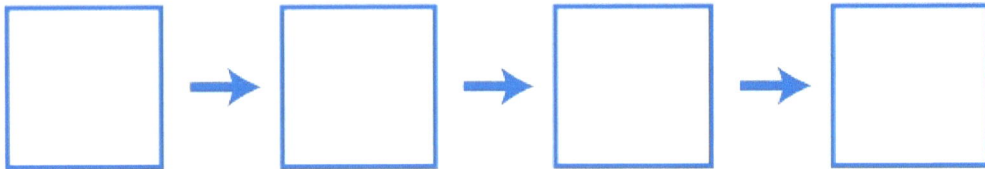

grizzly   **plankton**   **seal**   **cod**   **pigeon**   **orca**

Food chains exist between organisms that live in the same biome. The grizzly bear and pigeon do not live in the same biome as the rest of these organisms and so would not be a part of this food chain.

©2015 OnBoard Academics, Inc.   www.onboardacademics.com

Create a pond biome food chain.

Draw the food chain steps based on the illustration of the pond.

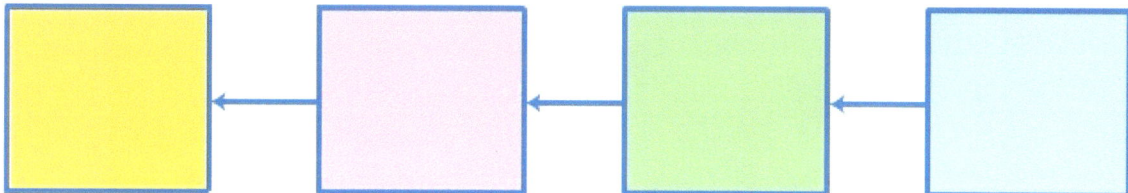

In most biomes, multiple food chains overlap and connect with each other. A network of all the food chains in a biome is called a food web.

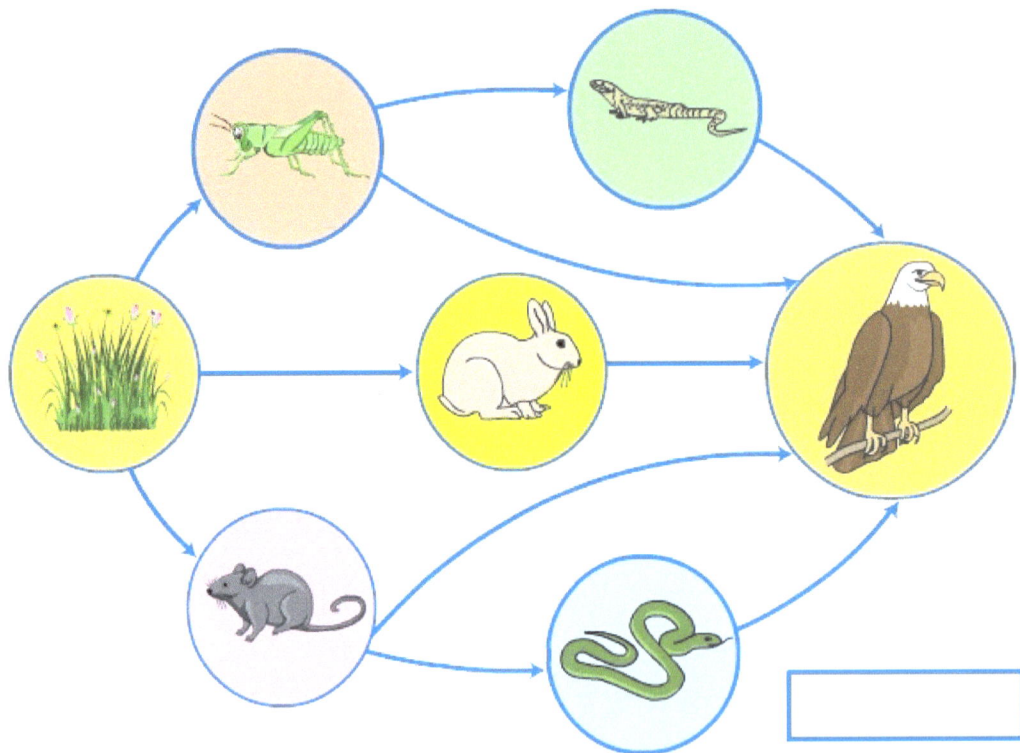

Food Web

In a food web, some living things can assume more than one role depending on the food chain within the web. This is why the eagle can be considered a secondary or tertiary consumer.

Based on the food web illustration below, place a check mark in the box to identify the roles of these organisms.

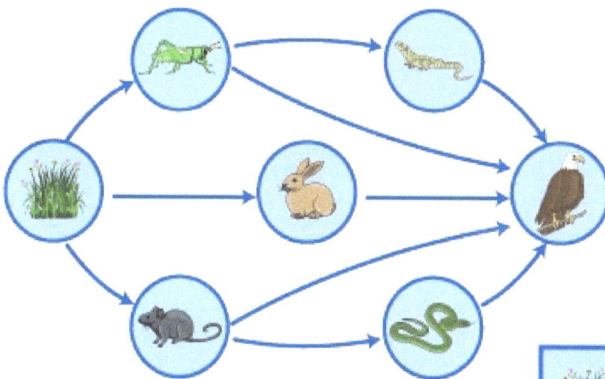

producer							
primary consumer							
secondary consumer							
tertiary consumer							

Build a food web by connecting these organisms.

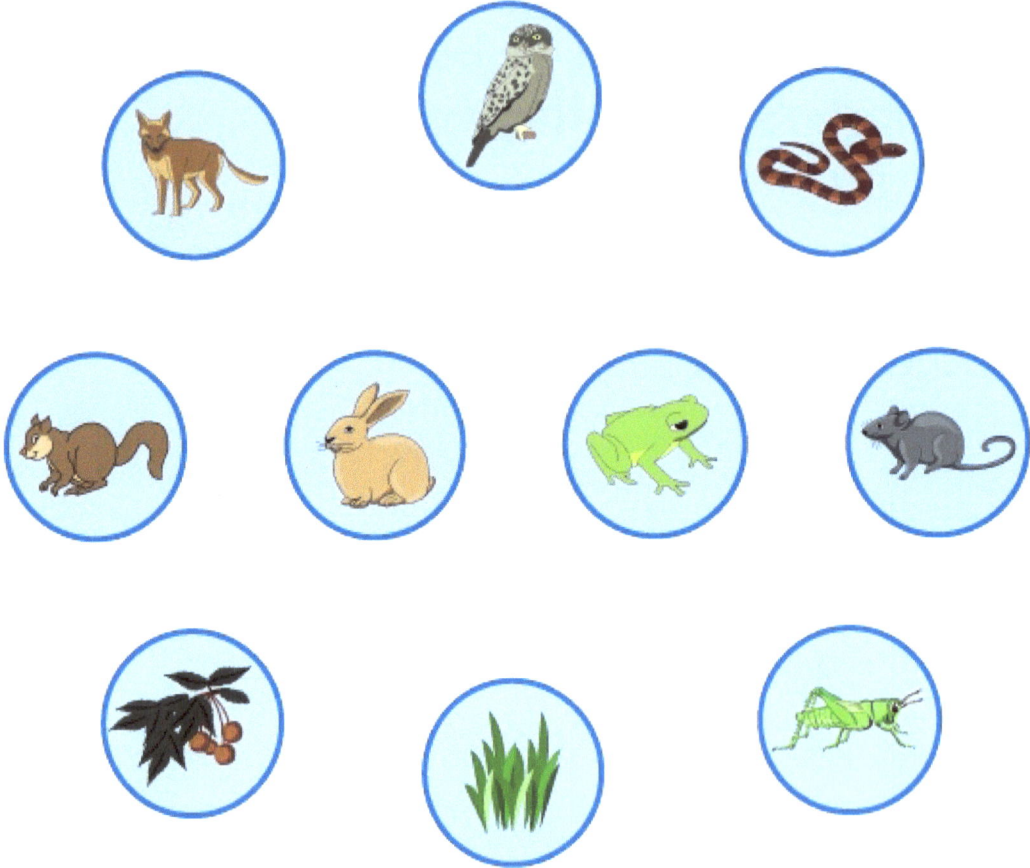

Scientists sometimes use other names to describe the roles of living things in a biome. An herbivore is a living thing that eats only plants. A carnivore eats only animals. Omnivores, such as humans, eat both plants and animals. A predator is an animal that eats another animal (its prey).

Use the label is the colored boxes to identify the different organisms.

The Importance of Balance in the Food Chain

Within the biome producers are the greatest in numbers because they are the basis of the food chain or food web. Similarly there are more primary consumers than secondary consumers and more secondary consumers than tertiary consumers. This is because each level of the food chain depends on the level below if for its energy. That's why the number of organisms in each level of a food chain look like a pyramid.

If a change occurs at any level of the food chain it can have a big effect on all other levels of the food chain. In our lake biome, imagine what would happen if the number of small fish were significantly reduced due to disease. The large fish would go hungry and would start to die out. This would then impact the bears because they wouldn't have enough food to eat.

Now lets see what would happen if we removed the bears for the food chain. The number of large fish would increase since there weren't enough bears to eat them. This would decrease the number of small fish because so many large fish would be feeding on the small fish.

> **The balance in a food chain is very delicate and a change at one level in the chain can affect the other levels... sometimes in disastrous ways.**

©2015 OnBoard Academics, Inc.   www.onboardacademics.com

Name_____

Food Chain Quiz

1.Place the following in order to form a food chain; snake, grass, chick, caterpillar.

_____ _____ _____ _____

2. In a biological community, an organism that eats the primary producer is what? Circle your answer.
   Producer,   Decomposer,   Consumer, None of these

3. A food chain always starts with plant life. True or False

4.What is a food web? Place a check mark next to the correct answer.
   a.   A network of food chains
   b. Provides a connection between trophic levels and a biological community
   c. Both a and b

5. Prey kill and eat other animals. True or false?

6. Organisms that eat both plants and animals are called what? Circle your answer.

   a.herbivores b. decomposers c. omnivores d. detrivors

Predators and Prey

©2015 OnBoard Academics, Inc.   www.onboardacademics.com

What is this owl most like to eat? Draw your answer next to the owl.

Predators and Prey

Predators are animals that hunt and eat other animals for food. There are many different predators throughout the world and they come in all shapes and sizes.

Prey are the animals that predators eat. This means that wherever there are predators there are prey. Like predators, prey come in all shapes and sizes.

©2015 OnBoard Academics, Inc.   www.onboardacademics.com

Match each predator with its most likely prey.

Animals often have many prey or many predators.

Many Prey   Many Predators

Many animals are both predators and prey.

Within a food chain, many animals have dual role as both predators and prey.

For example, in this food chain the frog is the predator because it eats the grasshopper.

However, the frog becomes the prey as soon as the snake enters the food chain.

©2015 OnBoard Academics, Inc.   www.onboardacademics.com

Complete the food chain.

Enter the initial for the animal that is both predator and prey in each food chain.

Balance of Predators and Prey

Predators and Prey help to keep
ecosystems in balance.

If we removed all of the prey
from an ecosystem then the
predators in that ecosystem
wouldn't have enough food and
eventually they would starve.

Similarly if we removed all
predators from an ecosystem
then the number of prey would
increase.

The extra prey would be
competing for a limited amount
of food. Eventually the food
would run out and the prey
would starve and die.

That's why having the right number of
predators and prey; having predators and
prey in balance, is vital to the health of all
living things within an ecosystem.

©2015 OnBoard Academics, Inc.   www.onboardacademics.com

Name: _____

Predators and Prey Quiz
Fill in the blank

Predators are animals that _____ other animals.
Prey are animals that are eaten by _____.
For example, an owl is a predator of a _____.
Some animals are _____ predators and prey.
For example, a robin might eat a _____, and
that same robin might get eaten by a _____.
Predators and prey keep ecosystems in _____.
They each have an important role. If there were no
_____, all the predators would starve to death.
If there were no predators, then the _____ of
the prey would increase so much that they would also
run out of _____.

food	population	balance	eat	both
hawk	grasshopper	predators	mouse	prey

How Plants Make Food

©2013 OnBoard Academics, Inc.

www.onboardacademics.com

Do you know your plant parts?

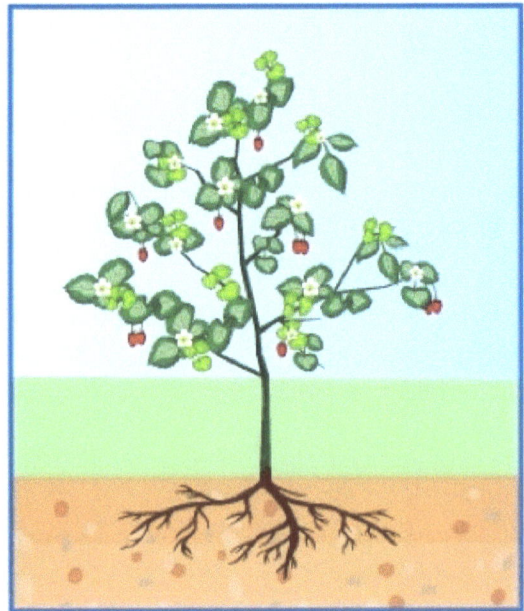

Connect the plant part with its name.

 **stem**    **leaf**    **flower**

   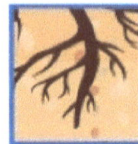  

 **fruit**    **seeds**    **roots**

©2013 OnBoard Academics, Inc.   www.onboardacademics.com

All of the food that we eat comes from plants or from animals that eat plats. But what do plants eat.

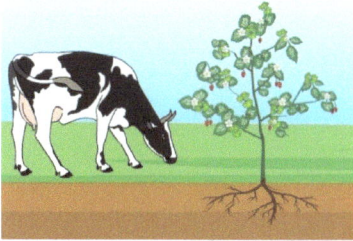

It might look like plants are just there doing nothing but unlike humans and other animals plants are busy making their own food. They do this using an amazing process called photosynthesis.

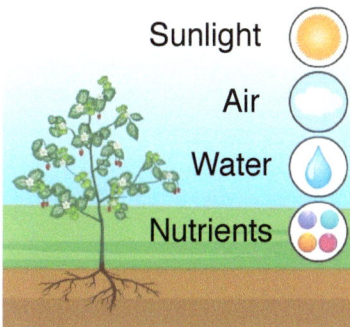

Sunlight
Air
Water
Nutrients

The four ingredients in the process of **photosynthesis** are sunlight, water, air and the nutrients from soil.

The main action happens in the plants leaves. The plant uses tiny openings in its leaves to capture sunlight and air.

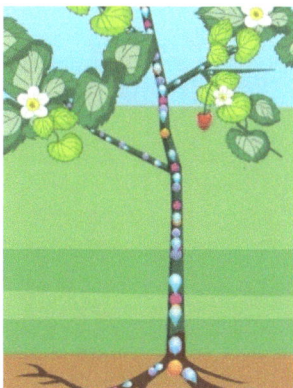

Water and soil nutrients enter through the opposite direction through the roots and then through the stems and on to the leaves. With all the ingredients in place, the leaves turn the sunlight, air, water and nutrients into sugar. This is the plant's food source that will help it to survive and to grow.

Arrange the steps to show how plants make food.

☐ + ☐ + ☐ + ☐ = ☐

| air | sugar | water | nutrients | sunlight |

Where do sunlight, air, water and nutrients enter the plant? Draw sunlight, air, nutrients, and water next to the plant part that they use to enter the plant.

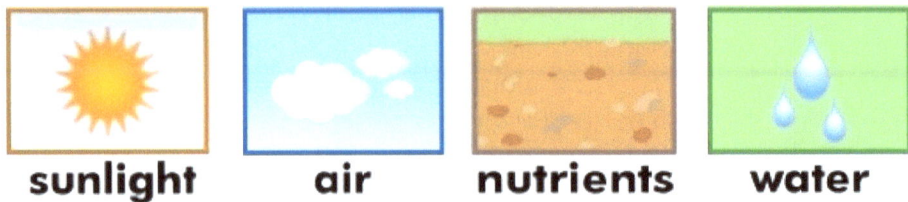

	flowers		leaves	
	leaves		roots	
	roots		stems	
	leaves		fruit	

| sunlight | air | nutrients | water |

©2013 OnBoard Academics, Inc.   www.onboardacademics.com

Name: _____

How Plants Make Food Quiz

1. Plants make their own food. True or false?

2. The process in which plants make their food is called
   _____.
   a. photosynthesis
   b. food synthesis
   c. Nutrient synthesis

3. The four main ingredients for photosynthesis to take place are sunlight, air, water and nutrients. True or false?

4. Water enters plants through leaves. True or false?

5. Plants turn sunlight, air, water and nutrients into ____.
   a. sugar
   b. salt
   c. fat

6. Air enters a plant through tiny openings in its _____.
   a. stem
   b. roots
   c. leaves

Living vs. Nonliving

Sort these objects.
Write the name of the object in the living or the non living box.

living	not living

Books

Tree

Dog

Computer

Backpack

Boy

Caterpillar

Pen

©2013 OnBoard Academics, Inc.   www.onboardacademics.com

Four Characteristics that Living Things Have in Common

Living things have four main things in common. All living things grow. For example humans grow from a baby to an adult.

While most plants grow from a seed to a plant.

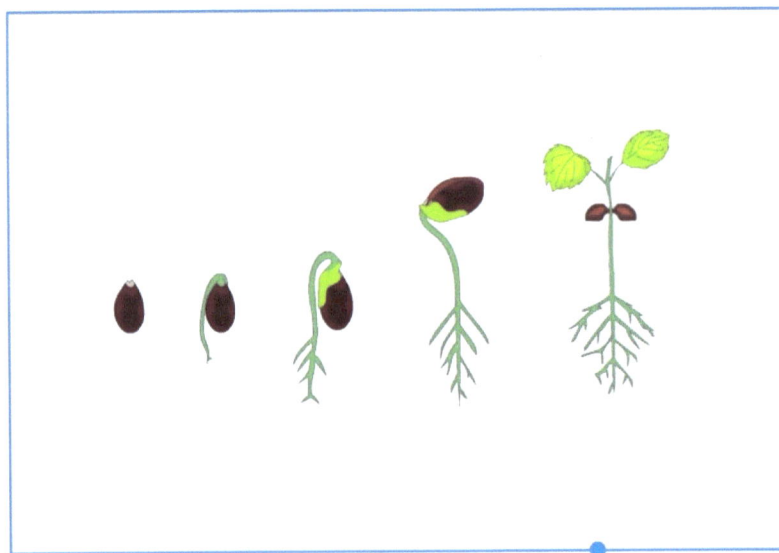

All living things need food and water to survive.   For example animals eat plants or animals or both.

Plants make their own food by using the sun's energy and the nutrients from the soil.

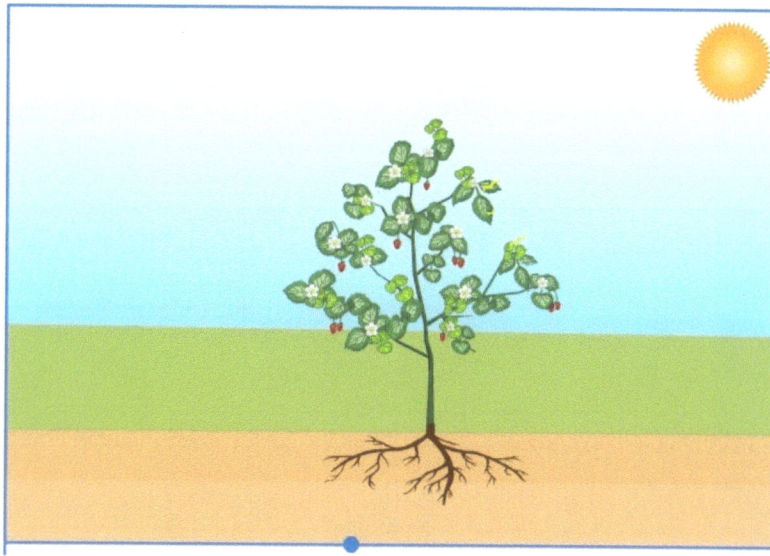

Both plants and animals need water to survive.

All living things have four main things in common: they grow, they need food and water, they reproduce, and they react to changes around them.

©2013 OnBoard Academics, Inc.   www.onboardacademics.com

All living things reproduce. This means that they have infants or babies. Some infants look much like their parents…

…while others look very different.

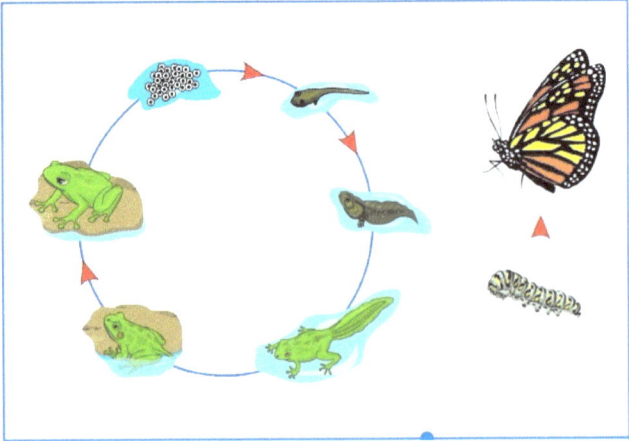

©2013 OnBoard Academics, Inc.   www.onboardacademics.com

Most plants reproduce from seeds.

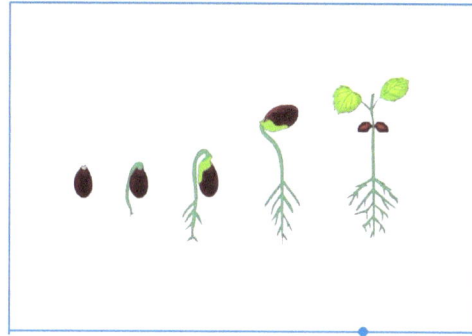

All living things react to change. For example the fur on the snowshoe hair changes from short and brown to thick and white during the winter.

The roots of a plant will grow toward the source of water.

©2013 OnBoard Academics, Inc.   www.onboardacademics.com

Which of these things grow?

Label each object.

√ grows   X doesn't grow   L living   N non-living

Which of these things needs food and water?
Label each item.

√ needs   X doesn't need   L living   N non-living

Which of these things react to changes around them?
Label each object.

√ reacts   X doesn't react   L living   N non-living

Which of these things reproduce?
Label each item.

√ does   X doesn't   L living   N non-living

©2013 OnBoard Academics, Inc.   www.onboardacademics.com

Living vs. Nonliving Quiz

All living things _____ over time.

All living things need _____ and _____.

All living things _____ to changes around them.

All living things _____; this
means that they have babies.

water   air   grow   plants   senses

reproduce   react   food   shrink